FOCUS ON

ELEMENTARY

ASTRONOMY

Teacher's Manual

3rd Edition

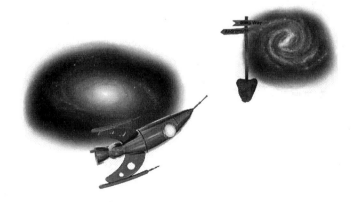

Rebecca W. Keller, PhD

Real Science-4-Kids

Focus On Elementary Astronomy Teacher's Manual— 3rd Edition
ISBN 978-1-941181-32-4

Published by Gravitas Publications Inc.
www.gravitaspublications.com
www.realscience4kids.com

GRAVITAS PUBLICATIONS

A Note From the Author

This curriculum is designed to provide an introduction to astronomy for students in the elementary level grades. *Focus On Elementary Astronomy—3rd Edition* is intended to be used as the first step in developing a framework for the study of real scientific concepts and terminology in astronomy. This *Teacher's Manual* will help you guide students through the series of experiments in the *Laboratory Notebook*. These experiments will help the students develop the skills needed for the first step in the scientific method — making good observations.

There are several sections in each chapter. The section called *Observe It* helps the students explore how to make good observations. The *Think About It* section provides questions for the students to think about and use to make further observations. In every chapter there is a *What Did You Discover?* section that gives the students an opportunity to summarize the observations they have made. A section called *Why?* provides a short explanation of what students may or may not have observed. And finally, in each chapter there is a section called *Just For Fun* that contains an additional activity.

The experiments take up to 1 hour. The materials needed for each experiment are listed on the next page and also at the beginning of each experiment.

Enjoy!

Rebecca W. Keller, PhD

Materials at a Glance

Experiment 1	Experiment 3	Experiment 5	Experiment 6	Experiment 7
clear night sky colored pencils	basketball ping-pong ball flashlight empty toilet paper tube glue or tape scissors marking pen a dark room	8 Styrofoam balls: Recommended (1) 10 cm (4 in) (1) 7.5 cm (3 in) (2) 5 cm (2 in) (2) 4 cm (1.5 in) (2) 2.5 cm (1 in) water-based craft paint: red, blue, green, orange, brown paintbrush water in a container misc. objects to represent planets (such as fruits, vegetables, candies, baking mixes) for *Just For Fun* section	colored pencils night sky daytime sky or textured surface **Optional** book or online information about constellations globe or basketball	Styrofoam ball pick, awl, or other thin, sharp object to poke a hole through the center of the ball nylon string scissors 2 or more marbles of different sizes cups that are different sizes

Experiment 2
colored pencils clear night sky basketball or other large object(s) **Telescope materials** empty cardboard paper towel tube 1-2 sheets of card stock or 1 manila file folder cut in half tape 2 lenses with different focal lengths*

Experiment 4
colored pencils night sky

Experiment 8	Experiment 9	Experiment 10	Experiment 11	Experiment 12
flashlight with new batteries glow sticks in assorted colors: may be found in places such as Walmart, toy stores, and online	student-selected materials to make a model of a galaxy, such as colored modeling clay, Styrofoam balls, tennis balls, marbles, sand, candies, etc. cardboard or poster board, .3-1 meter (1'-3') on each side **Optional** colored pencils or markers camera and printer	colored pencils a dark, moonless night sky far away from city lights **Optional** computer with internet access pictures of cities	2 bar magnets iron filings, purchased** or student collected shallow, flat-bottomed plastic container (or a plastic box top or large plastic jar lid) corn syrup plastic wrap Jell-O or other gelatin and items to make it assorted fruit cut in pieces and/or berries **Optional** cardboard box	small plastic pail that will fit in freezer water dirt small stones dry ice (available at most grocery stores) heavy gloves or oven mitts freezer If dry ice is in a block: safety goggles, mallet or hammer, grocery bag (cloth or paper)

* As of this writing, available from Home Science Tools: http://www.hometrainingtools.com   Item# OP-LEN4x15 and Item# OP-LEN4x50
Or: Look online for a telescope kit
** As of this writing, available from Home Science Tools: http://www.hometrainingtools.com   Item #CH-IRON

Materials
Quantities Needed for All Experiments

Equipment	Foods	Materials	Materials (cont.)
basketball basketball or other large object(s) cups, several - different sizes flashlight with new batteries freezer gloves, heavy, or oven mitts lenses (2) with different focal lengths* magnets, bar (2) marbles, 2 or more of different sizes pail, small plastic that will fit in freezer paintbrush pick, awl, or other thin, sharp object to poke a hole through the center of a Styrofoam ball ping-pong ball scissors **Optional** camera and printer computer with internet access globe plastic container, shallow, flat-bottomed (or a plastic box top or large plastic jar lid)	corn syrup fruit, assorted, cut in pieces, and/or berries Jell-O or other gelatin and items to make it objects, misc. to represent planets (such as fruits, vegetables, candies, baking mixes) for Just For Fun section	ball, Styrofoam (1) balls, 8 Styrofoam: Recommended (1) 10 cm (4 in) (1) 7.5 cm (3 in) (2) 5 cm (2 in) (2) 4 cm (1.5 in) (2) 2.5 cm (1 in) card stock, 1-2 sheets or 1 manila file folder cardboard or poster board, .3-1 meter (1'-3') on each side dirt dry ice (available at most grocery stores) If dry ice is in a block: safety goggles, mallet or hammer, bag, grocery (cloth or paper) glow sticks in assorted colors: may be found in places such as Walmart, toy stores, and online glue or tape iron filings, purchased** or student collected materials, misc student-selected to make a model of a galaxy, such as colored modeling clay, Styrofoam balls, tennis balls, marbles, sand, candies, etc.	paint, water-based craft: red, blue, green, orange, brown paper towel cardboard tube, empty pen, marking pencils, colored string, nylon stones, small tape toilet paper tube, empty water in a container **Optional** box, cardboard markers, colored pictures of cities plastic wrap

Other
room, darkened sky, dark, moonless, far away from city lights sky, daytime. or textured surface sky, night, clear **Optional** book or online information about constellations

* As of this writing, available from Home Science Tools: http://www.hometrainingtools.com   Item# OP-LEN4x15 and Item# OP-LEN4x50
Or: Look online for a telescope kit

** As of this writing, available from Home Science Tools: http://www.hometrainingtools.com   Item #CH-IRON

Contents

◇◇

Experiment 1

Observing the Stars

Materials Needed

- clear night sky
- colored pencils

Note: This experiment will take
6 days to complete.

Objectives

In this experiment, students will observe the stars, Moon, and planets on several different nights to determine their movement. Students will also explore how using the stars for navigation (celestial navigation) is possible.

The objectives of this lesson are:

- To observe changes in the position of stars, planets, and the Moon.
- To understand that it is possible to use the stars for navigation.

Experiment

I. Think About It

Read this section of the *Laboratory Notebook* with your students.

❶-❷ Have the students think about different ways they travel. Have them record their answers.

❸ Explain that *to navigate* means to plan and follow a route from one place to another. Have the students think about how a driver, pilot, or captain can find out which direction to travel. Discuss possible navigation tools such as:

- *compass*
- *GPS (Global Positioning System)*
- *radio navigation*
- *maps*

Have the students think about how using these tools enables modern people to travel to their destination. Have them record their answers.

❹ Help the students imagine what it would be like not to have any modern navigation tools.

Use the following questions to have the students explore the possibility of using the stars for navigation.

- *How easy or difficult would it be to use the stars to navigate?*
- *What happens when the sky is cloudy?*
- *Would traveling during the daylight be possible?*

II. Observe It

Read this section of the *Laboratory Notebook* with your students.

❶-❹ Have the students observe the night sky for six nights. Pick a single location (the backyard or front porch, for example), and observe the sky at the same time each night.

Help the students draw what they see. They do not have to draw every star. Try to help them find prominent stars and locate the same stars each night. Have them note if any of the stars have moved.

III. What Did You Discover?

Read the questions with your students.

❶-❹ Have the students answer the questions. These can be answered orally or in writing. There are no right answers, and their answers will depend on what they actually observed.

IV. Why?

Read this section of the *Laboratory Notebook* with your students.

Discuss any questions that might come up.

V. Just For Fun

Read this section of the *Laboratory Notebook* with your students.

Encourage the students to use their imagination to make up their own constellations.

Experiment 2

Building a Telescope

Materials Needed

- colored pencils
- clear night sky
- basketball or other large object(s)

Telescope materials*

- empty cardboard paper towel tube
- 1-2 sheets of card stock
  or
  1 manila file folder cut in half
- tape
- 2 lenses with different focal lengths
  Home Science Tools
  Item# OP-LEN4x15 and
  Item# OP-LEN4x50
  http://www.hometrainingtools.com

* Alternatively, you can look online for a telescope kit

Objectives

In this unit, students will build a simple telescope and explore how it can be used to make faraway objects appear larger with more detail.

The objectives of this lesson are to help students:

- Build a working telescope.
- Practice using a scientific tool and observe how using the tool changes what can be explored.

Experiment

I. Think About It

Read this section of the *Laboratory Notebook* with your students.

Gather all the materials needed for building the telescope.

Have the students take the sheet of card stock or 1/2 manila file folder and roll it into a tube with the longest side of the paper being the length of the tube. Have them tape it to hold it together. This, along with the cardboard paper towel tube, will make the barrel of the telescope.

Next, have the students examine the different pieces of the telescope.

Use the following questions to help them think about the various parts of the telescope.

- *Which items are the lenses?*
- *How many lenses are there? Why?*
- *Why are the lenses clear?*
- *Which items are the tubes?*
- *Why do you think there are different tubes?*
- *Where do you think the lenses will go?*

II. Observe It

Read this section of the *Laboratory Notebook* with your students.

Have the students assemble the telescope.

❶ Show them which lens is the eyepiece lens (shorter focal length) and have them tape this lens into one end of the card stock or file folder tube. They will need to either adjust the diameter of the tube they taped together previously, or if it seems easier, make a new tube. Have them tape the lens securely while being careful not to cover too much of the lens with tape.

❷ Have the students tape the objective lens, the longer focal length lens, to one end of the paper towel tube.

❸ Have them slide the open end of the heavy paper tube into the open end of the paper towel tube. This completes the telescope.

❹ Students are now to look at a faraway object, first using just their eyes and then by looking at it through the eyepiece of the telescope. The object they are viewing will appear upside down. Explain to the students that more complicated telescopes contain mirrors to turn the image right side up.

Have them experiment with sliding in and out the tube that has the objective lens. Help them observe that this will adjust the focus. If they are unable to focus the telescope, have them make one of the tubes longer by taping more card stock onto it.

❺ Students are to use the telescope to observe several objects that are far away. Have them pick objects that they can see while using only their eyes. Have them look at both small objects and large objects. Some suggestions:

- *a car parked across the street*

- *a building several blocks away*

- *a tree or flower at the far end of the yard or park*

- *a mountaintop or other geological feature*

Have the students compare the details they observe while using only their eyes and the details they observe when using the telescope. Have them note what kinds of details they can observe with the telescope that they cannot observe with their eyes only and draw their observations.

❻ Now that they have practiced with the telescope, have the students pick several stars, the Moon, or other features to observe in the night sky. Have them take their time observing through their telescope. Part of scientific discovery is slowing down and simply observing. Have them draw what they see.

III. What Did You Discover?

Read this section of the *Laboratory Notebook* with your students.

The questions can be answered verbally or in writing depending on the writing ability of the student. With these questions, help the students think about their observations. There are no "right" answers to these questions, and it is important for the students to write or discuss what they actually observed.

IV. Why?

Read this section of the *Laboratory Notebook* with your students.

Even when using scientific tools, the most important part of astronomy is observing. Good observing takes patience and practice. Ask the students to explain how using the telescope changes what they can observe.

Have the students imagine what it would be like to be Galileo. What would it have been like to be the first person to see planets in detail? How much more information about stars and planets can we explore today because of the telescope?

V. Just For Fun

A basketball is suggested for this experiment, but a different large object with a patterned or textured surface can be used. Help the students note as many features as possible about the object. Discuss with them any differences they notice when the object is held close, is far away and seen with only their eyes, and when it is far away and seen through the telescope. Ask which way they were able to get the most information. The least? Was there information they could not obtain when the object was far away?

The experiment may be repeated by moving the object different distances away or by using different objects.

Earth in Space

Materials Needed

- basketball
- ping-pong ball
- flashlight
- empty toilet paper tube
- glue or tape
- scissors
- marking pen
- a dark room

Objectives

In this experiment students will use simple materials to explore how light from the Sun affects the Earth and the Moon.

The objectives of this lesson are:

- To have the students observe how the Sun illuminates the Earth and the Moon.
- To demonstrate lunar and solar eclipses and the seasons.

Experiment

I. Observe It

In this section students will make a model Earth and use it to explore the illumination of the Earth and the Moon by the Sun.

Read this section of the *Laboratory Notebook* with your students.

❶-❷ Here the students will make a model of the Earth. Help them cut out the continents from the page and paste or tape them on the basketball in the appropriate positions, with North America, South America, and Greenland on one side; and Australia, Africa, Europe, Russia, and Asia on the other side. Have students refer to a globe if needed.

Have the students mark the approximate location of the North and South Poles by taping or gluing small pieces of paper to the basketball or by using a marking pen.

❸ Help your students cut the toilet paper cylinder 2.5 cm (one inch) from the end. This will create a ring to hold the basketball. Have them place the basketball on the ring and slightly tilt the ball so the North Pole is pointed slightly to the side. This represents the tilt of Earth's axis which is about 23° from vertical.

❹ A flashlight will be used to model the Sun. Have the students turn off the room lights and shine the flashlight on the basketball from some distance away. Have them observe how the flashlight illuminates the basketball.

❺ Students can place the flashlight on the floor, or you can hold the flashlight for them. They will now slowly rotate the basketball in a counterclockwise direction to simulate the rotation of Earth on its axis.

Guide your students' inquiry with the following questions:

- *Does the light cover the whole basketball or just one side?*
  just one side

- *If the light is shining on Asia, is North America light or dark?*
  dark

- *As you rotate the ball, does the light on Asia stay the same?*
  No, it changes.

- *Are Asia and Russia illuminated at the same time?*
  yes

- *Are Asia and North America illuminated at the same time?*
  no

- *If it is light in Asia, do you think it will be day or night in Russia?*
  day

- *If it is dark in Asia, do you think it will be day or night in North America?*
  day

- *Do you think that if the light is shining on South America it will be day or night in North America?*
  day

Have the students record their observations.

❻ Now the students will model the Moon. Have your students take the ping-pong ball and place it between the basketball and the flashlight, some distance away from the basketball. They will have to hold the ping-pong ball with their fingers. Have them observe the shadow the ping-pong ball casts on the basketball when it is between the basketball and the flashlight. This represents a solar eclipse during which the Moon blocks sunlight from reaching a portion of the Earth.

Have your students move the ping-pong ball in a circle around the basketball. As the ping-pong ball goes behind the basketball, the basketball casts a shadow on the ping-pong ball. This represents a lunar eclipse when Earth's shadow falls on the Moon.

Guide student inquiry with the following questions:

- *What happens to the basketball when the ping-pong ball is between the flashlight and the basketball?*
  [This represents the Moon between the Sun and the Earth—a solar eclipse.]
  The basketball will have a round shadow on it created by the ping-pong ball.

- *What happens when the basketball is between the flashlight and the ping-pong ball?*
  [This represents the Earth between the Sun and the Moon—a lunar eclipse.]
  The ping-pong ball is in the shadow created by the basketball.

Have the students record their observations.

II. Think About It

Read this section of the *Laboratory Notebook* with your students.

❶-❸ Have the students answer the questions. Encourage them to answer in their own words. Suggested answers are shown below.

> *(Answers may vary.)*
>
> ❶ Can you determine how day and night are created by the rotation of Earth?
>
> *As the Earth rotates, the Sun shines on different parts of the globe.*
>
> ❷ Can you observe how a lunar eclipse forms (where Earth casts a shadow on the Moon)?
>
> *When the Earth blocks the Sun's light from the Moon, the Moon has a shadow on it from the Earth.*
>
> ❸ Can you observe how a solar eclipse forms (where the Moon casts a shadow on Earth)?
>
> *When the Moon blocks the Sun's light from the Earth, the Moon's shadow falls on the Earth.*

❹ *Using the basketball and flashlight, can you show how the seasons are created? Explain how you would do this.*

Help your students model how the orbiting of Earth around the Sun creates seasons. Have the students stand some distance away from the flashlight Sun, holding the basketball Earth and tilting it slightly as it was in the *Observe It* section. One pole should be pointed toward the flashlight. Students will need to keep the tilt of the basketball constant as they circle the flashlight.

Have them hold the basketball and walk around the flashlight in a counterclockwise direction. To simulate the orbit of Earth around the Sun, students will remain facing in the same direction as they circle the flashlight. With the flashlight on their left and the "pole" of the basketball pointing toward it, they will start walking forward, then to the left, then backward, to the right, and forward again, completing the circle. Have them notice how one pole is tilted toward the Sun on one side of the circle and the other pole is tilted toward the Sun on the opposite side of the circle.

Discuss with the students how the parts of the Earth that are tilted toward the Sun receive more heat energy from the Sun and the parts tilted away receive less. At different times of the year different parts of the Earth are tilted more toward the Sun.

III. What Did You Discover?

Read this section of the *Laboratory Notebook* with the students.

❶-❹ Discuss the questions in this section with the students. Have them record their answers. Since these answers are based on what the students actually observed, their answers may vary.

IV. Why?

Read this section of the *Laboratory Notebook* with your students.

Review with the students how in this experiment they used a flashlight to represent the Sun, a basketball for Earth, and a ping-pong ball for the Moon. Explain that by doing this they built a model and that scientists build models to help them understand how things work. With this model the students were able to explore how the Earth rotates, creating day and night; how the movements and positions of the Moon and Earth create lunar and solar eclipses; and how seasons occur. (Students will learn more about making models in a following experiment.)

V. Just For Fun

Read this section of the *Laboratory Notebook* with your students.

Help the students look at the map on the basketball and mark the approximate location of where they live and the location of the equator. Help them orient the basketball and flashlight for each part of this experiment.

Encourage students to use their imagination as they experiment with the ideas presented in this section.

There are no right answers here. The purpose of this exercise is to encourage students to make observations, explore ideas, and use their imagination.

Seeing the Moon

Materials Needed

- colored pencils
- night sky

Note: This experiment will take two weeks to complete.

Objectives

In this experiment students will observe different phases of the Moon.

The objectives of this lesson are to have students:

- Practice making observations about a physical event that changes over time.
- Observe how their own observations may vary.

Experiment

I. Observe It

Read this section of the *Laboratory Notebook* with your students.

❶ Your students will be observing the movement, shape, and color of the Moon over fourteen nights. Pick a time during the evening to observe the Moon. It is recommended that the Moon be observed at the same time each night.

Guide your students' inquiry with the following questions.

- *What is the shape of the Moon?*
- *What is the color of the Moon?*
- *Can you observe any details in the Moon? Light or dark areas?*
- *Does the size of the Moon change?*

❷ Have the students record their observations each night.

II. Think About It

Read this section of the *Laboratory Notebook* with your students.

❶-❸ Have the students look at the various drawings or descriptions they have recorded. Have them use this information to answer the questions in as much detail as possible and in their own words.

III. What Did You Discover?

Read this section of the *Laboratory Notebook* with your students.

❶-❸ Discuss the questions in this section with the students. Have them record their answers. Answers may vary.

IV. Why?

Read this section of the *Laboratory Notebook* with your students.

Discuss with the students how the Moon appears to change shape, size, and color as it circles the Earth. Explain that the actual physical shape of the Moon does not change (the Moon itself doesn't actually grow larger or smaller or become full or half) but what changes is the way the Sun illuminates the Moon. Explain that as the Moon orbits the Earth, the Sun's light hits the Moon from different angles. The difference in how the Moon is illuminated by the Sun is what causes the Moon to appear to change shape.

Explain that when the Moon is closer to the horizon, it appears larger than when it is farther up in the sky. This happens because as the Moon nears the horizon, the atmosphere bends the light, making the Moon appear magnified. However, the physical size of the Moon has not changed.

V. Just For Fun

Read the text with your students.

Have your students observe any features of the Moon they find interesting. Can they see the "Man in the Moon?" Or "Jack and Jill?" Encourage them to use their imagination.

Have them draw what they observe and imagine. There are no right or wrong answers.

Experiment 5

Modeling the Planets

Materials Needed

- 8 Styrofoam craft balls
  Recommended sizes:
  - 1 – 10 cm (4 inch) ball
  - 1 – 7.5 cm (3 inch) ball
  - 2 – 5 cm (2 inch) balls
  - 2 – 4 cm (1.5 in) balls
  - 2 – 2.5 cm (1 in) balls
- Water-based craft paint
  Recommended colors:
  - red
  - blue
  - green
  - orange
  - brown
- paint brush
- water in a container
- *Just For Fun* section—miscellaneous objects to use to represent planets (such as fruits, vegetables, candies, baking mixes)

Objectives

In this experiment students will create a model of each of the planets.

The objectives of this lesson are for students to:

- Learn how to construct models.
- Explore the advantages and limitations of creating models.

Experiment

I. Observe It

Read this section of the *Laboratory Notebook* with your students.

❶-❷ In this section have the students look up the relative size and colors of each planet, using the illustrations in their *Student Textbook* or another reference.

Guide your students' inquiry with the following questions.

- *Of the eight planets, which one is the largest?*
  Jupiter

- *Which planet is the smallest?*
  Mercury

- *Are there any planets that are similar in size?*
  Earth and Venus, Neptune and Uranus

- *What color is Earth?*
  blue-green

- *What color is Mars?*
  reddish brown

- *What color is Neptune?*
  blue

- *What color is Jupiter?*
  brown stripes

Have the students record their observations about what each planet looks like.

❸ Have students choose a Styrofoam ball to represent each planet. They will need to compare the relative sizes of the planets to the various sizes of Styrofoam balls. Let your students decide on the sizes even if they are "wrong." A very important part of science is creating a model and discovering how well the model fits reality.

❹ Have the students paint the Styrofoam balls according to the information they've collected.

II. Think About It

Read this section of the *Laboratory Notebook* with your students.

❶-❸ Have the students think about how they chose the Styrofoam ball that would represent a certain planet. In general, see if they understand that for an accurate model the largest planet would need the largest Styrofoam ball, and the smallest planet would need the smallest Styrofoam ball. Discuss why this is important for building this model.

III. What Did You Discover?

Read this section of the *Laboratory Notebook* with your students.

❶-❸ Have the students discuss what they observed during the process of model building. Answers will vary.

IV. Why?

Read this section of the *Laboratory Notebook* with your students.

Discuss the process of model building. Model building is an important part of science. Models give scientists a way to visualize things they cannot observe directly. Models can also give scientists a deeper understanding of how natural processes work.

Discuss the limitations of model building. Models are only models and not "reality." Just because a scientist builds a model does not mean that it is an accurate representation of reality.

Scientists build physical models, like the students did in this project, but scientists also build mental models of how things work. Mathematical explanations and scientific concepts are part of the "models" scientists create to help them understand how the world works.

V. Just For Fun

Read this section of the *Laboratory Notebook* with your students.

Help your students create planet models from other materials. Edible planets can be made from fruits, vegetables, or candy. Planet models could also be baked using cupcakes or bread, or inedible objects could be used. Encourage your students to use their imagination.

The *Laboratory Workbook* has a space for students to list ideas for making planet models, and on the following page it has a space for students to draw the completed model.

Experiment 6

Tracking a Constellation

Materials Needed

- colored pencils
- night sky
- daytime sky or textured surface

Optional

- book or online information about constellations
- globe or basketball

Objectives

In this experiment students will observe a constellation of their choice to see whether its position changes in the night sky over the course of a week.

The objectives of this lesson are to have students:

- Make careful observations.
- Use their eyes as a tool in a scientific experiment.

Experiment

I. Think About It

Read this section of the *Laboratory Notebook* with your students.

Have the students answer the questions in this section. There are no right answers. Guide open inquiry with questions such as the following.

- *Do you think every star you can see is part of a constellation? Why or why not?*

- *Do you think there are lots and lots of constellations? Why or why not?*

- *Do you think people are coming up with new constellations all the time? Why or why not?*

- *If you look at the sky at different times of the night, do you think you will always see the stars in the same position? Why or why not?*

- *Do you think you would see the same groups of stars no matter where you are on Earth? Why or why not?*

- *From what location on Earth do you think you could see the most constellations? Why?*

II. Observe It

Read this section of the *Laboratory Notebook* with your students.

❶ Have the students pick a constellation to observe. To find more constellations than those mentioned in the *Student Textbook*, they can consult a book about constellations or look online for more information. Guide them in selecting a constellation that will be visible at the time they will be looking for it.

❷ Help the students locate the constellation they have chosen to observe. If they can't find it on the first night, have them try again the next night or several nights until they can see it.

❸ Have the students record the time and date when they first see their constellation. For the following six days they will view the constellation at the same time.

❹ A box is provided for students to record the position of the constellation by drawing or writing. Have them observe how high in the sky the constellation is and in which direction. It can be helpful to have them note a landmark to use to track the relative position of the constellation during the experiment—for example, how the constellations is positioned over a fence post, tree branch, or corner of a building.

❺ Have the students observe the constellation at the same time for six more days and record their observations about its location. If it's too cloudy to see the constellation, they can either note this for that day's observation or they can observe the constellation on six clear nights even if there are days in between the observations.

III. What Did You Discover?

Read this section of the *Laboratory Notebook* with your students.

Have the students answer the questions. Answers will be based on their observations.

IV. Why?

Read this section of the *Laboratory Notebook* with your students. Answer any questions that may come up.

If you have a globe, it can be used to demonstrate how the spin of Earth on its axis changes the view of the constellations during the course of a night. A globe can also be used show how the constellations that are visible at any one time varies according to one's location on Earth and how most or possibly all constellations will be visible from the equator. In addition, the globe can be used to show how Earth's orbit around the Sun changes Earth's position relative to the constellations and thus changes our view of the stars over the course of a year. A basketball or other ball may be used instead of a globe.

Although over time the stars do change their position in the universe relative to Earth, this happens so slowly that it isn't obvious in a lifetime.

V. Just For Fun

In this experiment students look at clouds and use their imagination to find shapes that remind them of some person, animal, or other object. Then they are asked to draw a picture of what they see and write a short story about it. The story can be one sentence or longer—wherever their imagination takes them. It can be fiction or nonfiction.

If there are no clouds, help students find a textured surface that provides enough variation to suggest different shapes.

Experiment 7

Modeling an Orbit

Materials Needed

- Styrofoam ball
- pick, awl, or other thin, sharp object to poke a hole through the center of the ball
- nylon string
- scissors
- 2 or more marbles of different sizes
- cups that are different sizes

Objectives

In this unit, students will observe how two opposing forces keep a Styrofoam ball in a circular orbit.

The objectives of this lesson are for students to:

- Create a model of a planetary orbit.
- Explore opposing forces.

Experiment

I. Think About It

Read this section of the *Laboratory Notebook* with your students.

Have the students think about what happens when a string is used to whirl a ball in the air. Use questions such as the following to guide their inquiry.

- *What do you think will happen when you hold the end of the string and whirl the ball in a circle? Will the ball stay in the same position on the string?*

- *Do you think the ball will move towards or away from your hand when you whirl it? Why?*

- *Do you think the ball will fly off the end of the string when you whirl it? Why or why not?*
  (This will happen if the knot isn't large enough.)

- *What will happen if you shorten the string? Will the ball move faster or slower with a shorter string?*

II. Observe It

Read this section of the *Laboratory Notebook* with your students.

❶ The students are to assemble the ball and string. Help them pierce the ball with a pick or sharp tool.

❷ Have the students tie a large knot at one end of the string and then thread the nylon string through the ball. When the ball and string are assembled, the ball should be near the unknotted end of the string with enough string at this end for the student to grasp firmly. The ball should be able to slide on the string, and the knot should be large enough that the ball won't come off the end of the string when it is whirled.

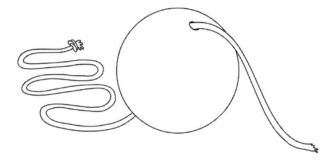

❸ It is useful to use the floor as a reference. Have the students whirl the ball around their hand with their hand fixed in the center of the circle. As they whirl it, the ball will slide outward along the string until it reaches the knotted end. The ball will then follow a circular path that stays at the same distance from their hand.

❹-❺ Have the students shorten and lengthen the string and observe how this changes the way the ball moves.

Encourage open inquiry with the following questions:

- *Does the ball go around faster or slower when the string is short?*

- *Is it easier or more difficult to spin the ball with a shorter string?*

- *If you slow down the speed at which you are spinning the ball, what happens to the ball?*

- *How fast can you spin the ball?*

III. What Did You Discover?

Read this section of the *Laboratory Notebook* with your students.

The questions can be answered verbally or in writing depending on the writing ability of the student. With these questions, help the students think about their observations. There are no "right" answers to these questions, and it is important for the students to write or discuss what they actually observed.

IV. Why?

Read this section of the *Laboratory Notebook* with your students.

There are two opposing forces that keep planets in a circular orbit around the Sun. One force, created by the speed and momentum of the planet, pushes the planet outward. The other force, the gravitational force of the Sun, pulls the planet inward. These two forces balance to keep the planets in circular orbits.

The balance of opposing forces is simulated in this experiment. As the ball is spun in a circle, a force causes it to travel outward. When it reaches the end of the string, the ball is pulled back towards the center, but since it is still rotating, it is also being pulled outward. The balance between the outward and inward forces keeps the ball in a circular orbit.

Planetary orbits are not quite circular, and the planets actually speed up as they near the Sun and slow down as they get farther away.

V. Just For Fun

Help the students use a marble in a cup to model an orbit. As the student moves the cup in a circular motion, the marble begins to circle the inside of the cup. In this experiment the cup creates an inward force on the marble while the outward force of the marble's momentum pushes against the cup. The two forces (inward and outward) are in balance which results in the marble circling the cup much as a planet orbits the Sun.

Have the students try using different size marbles, one at a time, in the same cup. Then they can try the marbles in different size cups. Help them notice any differences that may occur in the way the marbles move.

Brightest or Closest?

Materials Needed

- flashlight with new batteries
- glow sticks in assorted colors may be found in places such as Walmart, toy stores, and online

Objectives

The brightness of a star depends on how much light energy the star generates and not necessarily how close the star is to Earth. In this unit, students will observe the luminosity of two different light sources—a flashlight and a glow stick—to demonstrate that the brightest shining object may not be the closest.

The objectives of this lesson are for students to:

- Observe how different light sources have different luminosity (light energy output).
- Explore what happens when two light sources of different luminosity are different distances away.

Experiment

I. Think About It

Read this section of the *Laboratory Notebook* with your students.

Before the students perform the experiment, have them think about the differences between a glow stick and a flashlight. Draw on any previous knowledge they have about using flashlights or glow sticks.

Use the following questions to help guide the students' inquiry:

- *Which one (glow stick or flashlight) do you think is brighter?*

- *What happens to a flashlight when the batteries run down?*

- *How long do you think a glow stick will stay lit?*

- *If you put the glow stick and flashlight side-by-side and observe them from far away, do you think you would see both?*

- *What do you think will happen if the glow stick is closer to you? Will it look brighter than the flashlight?*

II. Observe It

Read this section of the *Laboratory Notebook* with your students.

❶ Help the students bend their glow stick so that the inner chamber cracks. This does not take much force. Too much force can cause the plastic to break, spilling the contents. Perform the experiment immediately. The glow stick lights up due to a chemical reaction and will only produce light for a few hours.

❷- ❹ Have the students observe how far into the distance the glow stick and the flashlight illuminate. This works best in a dark room or on a moonless, dark night. It is helpful to have the students use a point of reference. For example, you can have them stretch out their free hand and observe if they can see all their fingers in the light from the glow stick and then from the flashlight. Then have them look beyond the end of their hand to some object a little farther away and repeat the experiment with that object and so on until the light from the glow stick or flashlight is no longer bright enough to illuminate an object.

❺ Have the students place the glow stick and flashlight side-by-side on the ground or floor. Have the students walk about a meter (several feet) to several meters (yards) away from both and look toward the glow stick and flashlight without looking directly at the flashlight. If the flashlight is strong enough, it will likely wash out all of the luminosity from the glow stick. This simulates what happens when bright stars wash out the appearance of dim stars.

❻-❼ Have the students place the flashlight a meter or so (several feet) behind the glow stick and repeat their observations. The flashlight, being brighter, will appear closer than the glow stick.

Encourage open inquiry with the following questions.

- *Is the flashlight or the glow stick brighter?*

- *Why do you think the flashlight creates more light energy than the glow stick?*

- *How long do you think the glow stick's light will last?*

- *How long do you think the flashlight's light will last?*

III. What Did You Discover?

Read this section of the *Laboratory Notebook* with your students.

The questions can be answered verbally or in writing depending on the writing ability of the student. With these questions, help the students think about their observations. There are no "right" answers to these questions, and it is important for the students to write or discuss what they actually observed.

IV. Why?

Read this section of the *Laboratory Notebook* with your students.

Stars that appear closer may actually be farther away but brighter. By observing two different light sources of different luminosity, students can begin to understand how bright stars that appear closer or larger may in fact be farther away than smaller, dimmer stars because of their luminosity or light energy output.

Glow sticks produce light through a chemical process called chemiluminescence. When the inner chamber of a glow stick is broken, two chemicals are allowed to mix and react with each other. When this happens, fluorescent light is emitted. The luminosity of a glow stick is several times lower than that of flashlight.

V. Just For Fun

Have the students perform the experiment using different colored glow sticks and compare the results to their original experiment. They can also compare the glow sticks to each other to try to determine if one color is brighter than another. Ask them if they think that by combining the light from several glow sticks they will get a light that is as bright as the flashlight. Have them try it.

Experiment 9

Modeling a Galaxy

Materials Needed

- student-selected materials to make a model of a galaxy, such as colored modeling clay, Styrofoam balls, tennis balls, marbles, sand, candies, etc.
- cardboard or poster board, .3-1 meter (1-3 feet) on each side

Optional

- colored pencils or markers
- camera and printer

Objectives

In this experiment, students will explore model building and its limitations.

The objectives of this lesson are to have students explore how:

- Models help scientists ask questions.
- Scientists use different kinds of investigation, such as model building.

Experiment

I. Think About It

Read this section of the *Laboratory Notebook* with your students.

Explore open inquiry with questions such as the following:

- *How many houses do you think are in our neighborhood?*

- *How many neighborhoods do you think are within walking distance?*

- *How many neighborhoods have you visited?*

- *Do you think some neighborhoods look different from others? Why or why not?*

- *What else is in our neighborhood?*

- *Do you think a galaxy has neighborhoods? Why or why not?*

- *Do you think if you went from one part of a galaxy to another both parts would look the same? Why or why not?*

- *How many solar systems do you think are in a galaxy? Do all galaxies have the same number of solar systems? Why or why not?*

II. Observe It

Read this section of the *Laboratory Notebook* with your students.

❶ Have the students think about the different objects present in a galaxy and which ones they would like to represent in their model. Mentioned in the *Student Textbook* are solar systems (stars and planets), black holes, dust, and gases. Help the students select materials to make a model of a galaxy. Their galaxy can be as simple or as elaborate as they choose. The size of the model will be limited by the size of the cardboard or poster board provided as well as by the size of the various materials selected.

Use questions such as the following to help students select the materials to use for representing the various parts of the galaxy and to think about the model design.

- *Which parts of a galaxy do you want to have in your model? Why?*

- *How might you show different neighborhoods in your galaxy model?*

- *What material(s) would you use to represent the stars? Why?*

- *Do you want all the stars in your model to be the same size? Why or why or not?*

- *What material would you use to represent the planets? Why?*

- *Do you want all the planets in your galaxy model to be the same size and color? Why or why or not?*

- *What material might you use to represent dust in the galaxy?*

- *What do you think a black hole would look like in your model?*

Have the students list the objects they want to represent in their model and the materials they will use.

❷ Provide the cardboard or poster board for the students and have them spread out their model making materials next to it or along an edge.

❸ Guide the students in designing and drawing their galaxy model by having them think about where the solar systems and any other objects might go, how many of them they will create, and how far apart and how big they will be. Have them think about the placement of different objects in terms of creating neighborhoods. There are no right answers and they can add as many solar systems and other objects as they like. Help them notice the limitations of their model.

❹ Have the students build the model galaxy, referring to the design they created. Have them construct at least one solar system with at least one Sun and 3 or more planets. Several smaller solar systems can be constructed or drawn on the cardboard. Solar systems may be any size. Have them add any other objects they want to represent.

❺ Have the students record any features of their model that they think are unique or unexpected. Or they might record the features they like best.

❻ Help the students photograph their model and print a copy to tape in the *Laboratory Notebook,* or they can draw it or describe it in writing.

III. What Did You Discover?

Read this section of the *Laboratory Notebook* with your students.

Have the students answer the questions. There are no right answers and their answers will depend on what they actually observed..

IV. Why?

Read this section of the *Laboratory Notebook* with your students.

Discuss any questions that might come up.

V. Just For Fun

Encourage students to use their imagination freely to think about what it might be like to go to the end of the universe. Do they think there's an end? If so, what might it look like? What could be seen along the way there? Have them record their ideas by drawing and/or writing in the space provided.

Experiment 10

See the Milky Way

Materials Needed

- colored pencils
- a dark, moonless night sky far away from city lights

Optional

- computer with internet access
- pictures of cities

Objectives

In this unit, students will use the model of a city to help them think about and observe the Milky Way Galaxy.

The objectives of this lesson are for students to:

- Observe a city to see that it typically has more buildings at its center than it does at the edges.
- Compare how observations made about a city's structure and organization can be used to better understand that of a galaxy.

Experiment

I. Think About It

Read this section of the *Laboratory Notebook* with your students.

Have students think about the density of buildings in the center of a city compared to their density on the edges of a city. Most cities have a greater number of close together buildings in the center and fewer, more widely spaced buildings on the outskirts. Have students think about how their observations of a city's structure can be used to model the structure of a galaxy.

Encourage open inquiry with questions such as the following.

- *How many buildings are in the downtown section of a city?*
- *How many buildings are at the edges of a city?*
- *Are the buildings in the downtown area typically taller/bigger or shorter/smaller than the buildings at the edges?*
- *Are the buildings in the downtown area typically closer together or farther apart than the buildings at the edges?*
- *How might the center of a galaxy like ours be compared to the center of a city?*
- *How might the edges of a city or the suburbs be compared to the edges of a galaxy?*
- *Do all cities have the same shape? Do all galaxies have the same shape? Why or why not?*
- *Do you think cities stay the same size? Do you think galaxies stay the same size? Why or why not?*

II. Observe It

Read this section of the *Laboratory Notebook* with your students.

To observe the Milky Way, choose a clear, moonless night far away from city lights. Allow some time for students' eyes to adjust to the dark and have them look with their eyes only.

For students in the Northern Hemisphere, the best time to view the Milky Way is in the late summer and early winter when the Milky Way will look brighter. In the Southern Hemisphere winter is the best time for viewing.

**Best times and dates
for viewing the Milky Way in the Northern Hemisphere**

Summer Milky Way Times & Dates		*Winter Milky Way Times & Dates*	
8 PM	October 14	8 PM	February 10
9 PM	September 29	9 PM	January 25
10 PM	September 14	10 PM	January 10
11 PM	August 30	11 PM	December 26
12 AM	August 15	12 AM	December 12
1 AM	July 31	1 AM	November 26
2 AM	July 16	2 AM	November 11

**Best times and dates
for viewing the Milky Way in the Southern Hemisphere**

Summer Milky Way Times & Dates		*Winter Milky Way Dates & Times*	
8 PM	April 8	8 PM	August 28
9 PM	March 24	9 PM	August 12
10 PM	March 9	10 PM	July 28
11 PM	February 21	11 PM	July 12
12 AM	February 5	12 AM	June 27
1 AM	January 22	1 AM	June 13
2 AM	January 7	2 AM	May 29

III. What Did You Discover?

Read this section of the *Laboratory Notebook* with your students.

Help students think about their observations while answering these questions. There are no "right" answers to these questions, and it is important for the students to write or discuss what they actually observed.

IV. Why?

Read this section of the *Laboratory Notebook* with your students.

The Milky Way Galaxy contains a vast number of stars. The majority of stars in the galaxy reside in a flat disk, and there is a concentration of stars near the center. Our solar system is located within the flat disk of stars. Because Earth is located toward the outer edge of the galaxy, we can see a multitude of stars as we look through the galaxy toward the center. The band of light and stars we call the Milky Way appears as a band because we are looking edge-on through the disk of stars toward the galaxy center.

Artist's Rendition of Milky Way Galaxy as seen from outside and looking toward the center   Credit: NASA/UMass/Caltech

V. Just For Fun

Help your students set up Google Earth:

❶ Go to http://earth.google.com and click "Download Google Earth."

❷ Click "Agree and Download."

❸ Once the file has been downloaded, follow the directions to install the program.

❹ Open the Google Earth program on your computer.

❺ Set up Google Earth in Sky mode by clicking on the planet symbol in the top menu bar and selecting "Sky" from the drop down menu. Or, at the top of the page, click "View," then "Explore," and then select"Sky."

❻ Type "Milky Way" into the search box. Once the Milky Way appears, you can zoom in and out using the +/- bar.

Students can also go to Google Images (www.google.com/imghp) and search on Milky Way Galaxy to find some beautiful photos and illustrations of the galaxy.

If your students are interested, you can have them use Google Earth to take a look at some of the planets.

How Do Galaxies Get Their Shape?

Materials Needed

- 2 bar magnets
- iron filings, purchased* or student collected
- shallow, flat-bottomed plastic container (or a plastic box top or large plastic jar lid)
- corn syrup
- plastic wrap
- Jell-O or other gelatin and items needed to make it
- assorted fruit cut in pieces and/or berries

Optional

- cardboard box

* As of this writing, available from Home Science Tools: http://www.hometrainingtools.com Item #CH-IRON

Objectives

In this unit students will model the forces that help shape galaxies.

The objectives of this lesson are for students to:

- Use model building as a way to explore elements of science that can't be directly tested.
- Observe how forces move objects.

Experiment

Students can gather iron filings themselves by putting a magnet in a plastic bag and dragging the bag through some dirt. Iron that is in the dirt will collect on the outside of the bag. Place the bag containing the magnet inside another plastic bag and then remove the magnet from the inner bag. The iron filings will fall into the outer bag. Repeat several times until about 5 ml (1 teaspoon) of iron filings has been collected.

Alternatively, iron filings can be purchased.

I. Think About It

Read this section of the *Laboratory Notebook* with your students.

Have the students think about how galaxies might form. Encourage open inquiry with questions such as the following:

- *How do you think a galaxy might form?*
- *What do you think holds the stars and planets together?*
- *Do you think galaxies can have any kind of shape? Why or why not?*
- *Do you think galaxies can change their shape? Why or why not?*
- *What do you think might happen if two galaxies collided?*

II. Observe It

Read this section of the *Laboratory Notebook* with your students.

❶ Provide the students with a shallow, flat-bottomed plastic container, and help them pour corn syrup into it until the syrup is just below the rim of the container.

❷ Help the students pour iron filings into the corn syrup. Iron filings can irritate eyes and lungs, so they should be handled carefully.

❸ The students should now cover the container with plastic wrap to prevent spilling.

❹ The container can be placed on a table that has a top thin enough for the magnets to work when they are being moved under it; or you can place a cardboard box on its side with the open end toward the student and the container on top; or the container can be held for the student. There should be enough space under the container for both of the student's hands to fit and be able to move.

Have students place the magnets underneath the plastic container and observe the movement of the iron filings. The corn syrup will create drag on the iron filings, slowing the movement. Encourage students to be patient while observing the filings.

❺ Have the students take one of the magnets and create a swirling pattern with it. The swirling motion will simulate how a spiral galaxy might form. Have the students notice that the source of the force (the magnet) moves and that this creates movement in the iron filings.

❻ Now have them take a magnet in each hand and create opposite swirling patterns.

❼ Have the students bring the magnets together to see what happens.

❽ Allow the students some time to "play" with the magnets and iron filings. Slowing down to "play" with science is an essential part of scientific inquiry. Encourage the students to bring the magnets together, pull them apart, swirl them together, swirl them separately, swirl them in opposite directions, and so on. How many different shapes can they create?

III. What Did You Discover?

Read this section of the *Laboratory Notebook* with your students.

In answering these questions, help the students think about their observations. There are no "right" answers to these questions, and it is important for students to write or discuss what they actually observed.

IV. Why?

Read this section of the *Laboratory Notebook* with your students.

Gravity is a force that acts between any two objects that have mass, pulling them toward each other. Magnetism depends on the arrangement of electrons in objects and can either pull objects toward each other or repel them, moving them away from each other. All objects are affected by gravity, but not all objects are affected by magnetism.

Even though magnetic forces are different from gravitational forces, in this experiment the magnetic action is pulling on the iron filings so the force appears to be like that of gravity. Therefore, magnets can be used to simulate how gravity works to pull and shape galaxies. The magnetic forces act on the iron filings, and when the magnets are moved, the iron filings are pulled along. In a similar way stars pull on smaller objects such as planets, comets, and asteroids. Stars also pull on neighboring stars and together they shape a galaxy.

V. Just For Fun

Help your students set up a Jell-O galaxy. Any type of fruit can be used to represent planets and stars. Help your students cut the fruit into small pieces and arrange the fruit or swirl it in the Jell-O mixture to create an elliptical galaxy, a spiral galaxy, or an irregular galaxy. Encourage them to think about whether or not other shapes of galaxies might be possible.

Making a Comet

Materials Needed

- small plastic pail that will fit in freezer
- water
- dirt
- small stones
- dry ice (available at most grocery stores)
- heavy gloves or oven mitts
- freezer

If dry ice comes in a block:

- safety goggles
- mallet or hammer
- cloth or paper grocery bag

Objectives

In this unit, students will model a comet.

The objectives of this lesson are for students to:

- Observe how a mixture of water, dirt, and rocks can be frozen to form a model comet.
- Explore how changing the ratio of water:dirt:rocks changes a comet.

Experiment

I. Think About It

Read this section of the *Laboratory Notebook* with your students.

Encourage open inquiry with the following questions.

- *How big do you think a real comet is? Why?*

- *Do you think a real comet is mostly ice or mostly dirt or mostly rock?*

- *How much ice do you think is needed to hold rocks and dirt together?*

- *Do you think a real comet breaks apart, or does it vaporize as it nears the Sun? Could it do both? Why or why not?*

II. Observe It

Read this section of the *Laboratory Notebook* with your students.

In this experiment students will make a model comet from water, dirt and rocks. Help the students make careful observations about the comet they are creating.

❶ Have students collect some dirt and small stones.

❷ Have students pour dirt and stones into a small pail and add enough water to cover, leaving several inches between the top of the pail and the water. As the water freezes it will expand.

❸ Place the pail in the freezer and allow enough time for the water to freeze completely.

❹ Have the students tap the frozen comet model out of the pail. They may need to run a little warm water over the outside of the pail to get the frozen mixture to release.

❺ Encourage the students to observe the frozen comet model carefully and draw or write about any details they notice.

❻ Students will now observe their comet model as it melts. Guide them in making their observations. Do the rocks fall away in large chunks? What happens to the dirt? What else do they notice? Have them record their observations.

❼-❽ Have the students repeat the experiment first with more water and then more dirt. By carefully recording their observations, they will be able to compare how a comet containing lots of ice acts compared to a comet made mostly of dirt.

III. What Did You Discover?

Read this section of the *Laboratory Notebook* with your students.

In answering these questions, help the students think about their observations. There are no right answers to these questions, and it is important for the students to write or discuss what they actually observed.

IV. Why?

Read this section of the *Laboratory Notebook* with your students.

A real comet is a large chunk of ice and rock that might look similar to the model comet the students have created. However, the ice in real comets vaporizes rather than melts. This experiment demonstrates what happens to a comet as it loses its ice, although it does not replicate the exact method by which the ice is lost. Explain to the students that astronomers performing experiments are not always able to duplicate the exact conditions that exist in space.

Comets also contain frozen gases such as carbon dioxide and carbon monoxide. The tails of comets will be different colors depending on the kinds of gases that are vaporizing from the comet.

V. Just For Fun

Using dry ice will create a more realistic comet. Dry ice is made of frozen carbon dioxide, and it vaporizes rather than melts. This is fun for the students to watch.

Dry ice is very cold and can burn the skin with minimal contact. Please **use extreme caution** when building a comet with dry ice, and don't let bare skin touch the dry ice. Wear heavy gloves or oven mitts.

If the dry ice is in a solid block, it will need to be broken into small pieces. You can place it in a cloth or paper grocery bag and hit it with a mallet or hammer until it is broken up. (You may want to wear goggles during this process.) Then the pieces can be added to the water, dirt, rock mixture.

More REAL SCIENCE-4-KIDS Books
by Rebecca W. Keller, PhD

Building Blocks Series yearlong study program — each Student Textbook has accompanying Laboratory Notebook, Teacher's Manual, Lesson Plan, Study Notebook, Quizzes, and Graphics Package

Exploring Science Book K (Activity Book)
Exploring Science Book 1
Exploring Science Book 2
Exploring Science Book 3
Exploring Science Book 4
Exploring Science Book 5
Exploring Science Book 6
Exploring Science Book 7
Exploring Science Book 8

Focus On Series unit study program — each title has a Student Textbook with accompanying Laboratory Notebook, Teacher's Manual, Lesson Plan, Study Notebook, Quizzes, and Graphics Package

Focus On Elementary Chemistry
Focus On Elementary Biology
Focus On Elementary Physics
Focus On Elementary Geology
Focus On Elementary Astronomy

Focus On Middle School Chemistry
Focus On Middle School Biology
Focus On Middle School Physics
Focus On Middle School Geology
Focus On Middle School Astronomy

Focus On High School Chemistry

Super Simple Science Experiments

21 Super Simple Chemistry Experiments
21 Super Simple Biology Experiments
21 Super Simple Physics Experiments
21 Super Simple Geology Experiments
21 Super Simple Astronomy Experiments
101 Super Simple Science Experiments

Note: A few titles may still be in production.

Gravitas Publications Inc.
www.gravitaspublications.com
www.realscience4kids.com